Endangered Species

Understanding Words in Context

Curriculum Consultant: JoAnne Buggey, Ph.D.
College of Education, University of Minnesota

By Robert Anderson

Greenhaven Press, Inc.
Post Office Box 289009
San Diego, CA 92198-9009

Titles in the opposing viewpoints juniors series:

Advertising	Male/Female Roles
AIDS	Nuclear Power
Alcohol	The Palestinian Conflict
Animal Rights	Patriotism
Causes of Crime	Population
Child Abuse	Poverty
Death Penalty	Prisons
Drugs and Sports	Smoking
Endangered Species	Television
The Environment	Toxic Wastes
Garbage	The U.S. Constitution
Gun Control	The War on Drugs
The Homeless	Working Mothers
Immigration	Zoos

Cover photo by: FPG International/Telegraph Colour Library

Library of Congress Cataloging-in-Publication Data

Anderson, Bob, 1950-
    Endangered species : understanding words in context / by Robert Anderson; curriculum consultant, JoAnne Buggey.
      p. cm. — (Opposing viewpoints juniors)
    Summary: Presents differing views on four issues related to endangered species, such as should endangered species be saved and does hunting threaten certain species. Includes critical thinking activities.
  ISBN 0-89908-608-X
    1. Endangered species—Juvenile literature. 2. Wildlife conservation—Juvenile literature. 3. Vocabulary—Juvenile literature. 4. Critical thinking—Juvenile literature. [1. Rare animals. 2. Wildlife conservation. 3. Critical thinking. 4. Vocabulary.] I. Buggey, JoAnne. II. Title. III. Series.
QL83.A83   1991
591.52'9—dc20   91-29890
                                                                                       CIP
                                                                                                 AC

No part of this book may be reproduced or used in any other form or by any other means, electrical, mechanical, or otherwise, including, but not limited to photocopy, recording, or any information storage and retrieval system, without prior written permission from the publisher.

Copyright 1991 by Greenhaven Press, Inc.

CONTENTS

The Purpose of This Book: An Introduction to Opposing Viewpoints 4
Skill Introduction: Understanding Words in Context 5
Sample Viewpoint A: People should help save dolphins 6
Sample Viewpoint B: People should not help save dolphins 7
Analyzing the
Sample Viewpoints: Understanding Words in Context 8

Chapter 1

Preface: Should Endangered Species Be Saved? 9
Viewpoint 1: Endangered species should be saved 10
Viewpoint 2: Endangered species should not be saved 12
Critical Thinking Skill 1: Understanding Words in Context 14

Chapter 2

Preface: Do Humans Cause the Extinction of Other Species? 15
Viewpoint 3: Humans cause the extinction of other species 16
Viewpoint 4: Humans prevent the extinction of other species 18
Critical Thinking Skill 2: Using New Words in Sentences 20

Chapter 3

Preface: Does Hunting Threaten Species? 21
Viewpoint 5: Hunting threatens species 22
Viewpoint 6: Hunting saves species 24
Critical Thinking Skill 3: Recognizing and Defining Cognates 26

Chapter 4

Preface: How Can Elephants Be Saved from Extinction? 27
Viewpoint 7: Banning the sale of ivory protects elephants 28
Viewpoint 8: Banning the sale of ivory endangers elephants 30
Critical Thinking Skill 4: Reading Comprehension 32

THE PURPOSE OF THIS BOOK

An Introduction to Opposing Viewpoints

When people disagree, it is hard to figure out who is right. You may decide one person is right just because the person is your friend or relative. But this is not a very good reason to agree or disagree with someone. It is better if you try to understand why these people disagree. On what main points do they differ? Read or listen to each person's argument carefully. Separate the facts and opinions that each person presents. Finally, decide which argument best matches what you think. This process, examining an argument without emotion, is part of what critical thinking is all about.

This is not easy. Many things make it hard to understand and form opinions. People's values, ages, and experiences all influence the way they think. This is why learning to read and think critically is an invaluable skill.

Opposing Viewpoints Juniors books will help you learn and practice skills to improve your ability to read critically. By reading opposing views on an issue, you will become familiar with methods people use to attempt to convince you that their point of view is right. And you will learn to separate the authors' opinions from the facts they present.

Each Opposing Viewpoints Juniors book focuses on one critical thinking skill that will help you judge the views presented. Some of these skills are telling fact from opinion, recognizing propaganda techniques, and locating and analyzing the main idea. These skills will allow you to examine opposing viewpoints more easily. The viewpoints are placed in a running debate and are always placed with the pro view first.

SKILL INTRODUCTION

Understanding Words in Context

Whenever you read, you may come across words you do not understand. Sometimes, because you do not know a word or words, you will not fully understand what you are reading. One way to avoid this is to interrupt your reading and look up the unfamiliar word in the dictionary. Another way is to examine the unfamiliar word in context, or to study the words, ideas, and attitudes that surround the unfamiliar word.

In this Opposing Viewpoints Juniors book, you will be asked to determine the meaning of words you do not understand by considering their use in context.

Sometimes a word that has the same meaning as the unfamiliar word will be used in the sentence or in a surrounding sentence:

> Many animal species are **endangered** by human activities. Their lives are threatened by people destroying the environment.

The unfamiliar word is **endangered**. The clue is the word *threatened*. The second sentence is relating the same idea as the first sentence, but it is a little more specific. So, the words threatened and endangered should mean about the same thing. In fact, they do.

Often, the surrounding sentences will not contain a word similar to the unfamiliar word. They may, however, contain ideas that suggest the meaning of the unknown word:

> The United States has many **assets**. It has beautiful scenery, natural resources, generous people, and great wealth.

The meaning of the word **assets** can be determined by studying the ideas around it. Beautiful scenery, natural resources, generous people, and great wealth are all clues to the meaning of the word *assets*, things that are desirable to own.

In some cases, opposite words and ideas can offer the reader clues about an unfamiliar word:

> Seldom will you find a man with more than one wife. Most men are **monogamous**.

The unfamiliar word is **monogamous**. The clue is *more than one wife* in the first sentence. In this case, the author is saying, "You do not usually find a man with more than one wife, so you do usually find a man with only one wife." Monogamous, then, must mean "having only one wife."

Sometimes the meanings of unfamiliar words are more difficult to determine. You just have to pay attention to the meaning of the sentence to figure them out:

> Pickpockets can be **incapacitated** by cutting off their hands.

To determine what **incapacitated** means, you have to figure out that cutting off pickpockets' hands would stop them from picking people's pockets. So incapacitated must mean that a person cannot do something.

Sometimes you will not be able to determine what a word means by its context. You will have to look it up in a dictionary.

In the following viewpoints, several of the words are highlighted. You must determine the meanings of these highlighted words by studying them in context. As you read the material presented in this book, stop at the unfamiliar words and try to determine their meanings. Finally, use a dictionary to see how well you have understood the words.

We invited two students to give their opinions on the endangered species issue. We asked each of them to look up an unfamiliar word and use it in a sentence. In the students' viewpoints and the ones on the following pages, try to figure out the meanings of words you do not understand.

SAMPLE VIEWPOINT A *Kioko:*

People should help save dolphins.

Dolphins are beautiful and intelligent creatures. I think it is a crime that the tuna-fishing industry kills hundreds of them every year in their fishing nets. The workers on the fishing boats don't even seem to care that they catch dolphins in their nets along with tuna. They just let the wounded dolphins die.

I saw a news story about this just the other day. Some people were protesting at a harbor in southern California where the tuna boats dock. Some of the protesters had signs saying not to buy tuna fish. My mom said that if people refuse to buy tuna, the tuna companies will lose lots of money. They will be forced to change their fishing methods to keep the dolphins out of the tuna nets. My family has joined the **boycott** of tuna. I feel good that we are helping save the dolphins. I think everybody should help save these gentle creatures.

SAMPLE VIEWPOINT B
Manuel:

> People should not help save dolphins.

I like dolphins as much as everyone else does. But I don't think they are more important than human beings. It is too bad that a few dolphins get caught in fishing nets intended for tuna. But to insist that the tuna boats change their methods of fishing is unreasonable. My brother works on a tuna-fishing boat in California. He says that such a change would cost so much money that there would be no profit in fishing for tuna. The tuna companies would go out of business. Thousands of people would lose their jobs. Their families would suffer too. It is not fair to **jeopardize** the lives and jobs of so many people just to save a few dolphins. I believe that people who are trying to save the dolphins by hurting the tuna-fishing industry are mixed up. They don't seem to realize that people are more important than dolphins.

ANALYZING THE SAMPLE VIEWPOINTS

Manuel and Kioko have very different opinions about endangered dolphins. Both of them used words you might not understand. But you should be able to discover the meanings of these words by considering their use in context.

Kioko:

My mom said that if people refuse to buy tuna, the tuna companies will lose lots of money. . . . My family has joined the *boycott* of tuna.

NEW WORD	CLUE WORD OR IDEA	DEFINITION
boycott	refuse to buy	to refuse to deal with someone or something

Manuel:

Thousands of people would lose their jobs. Their families would suffer too. It is not fair to *jeopardize* the lives and jobs of so many people just to save a few dolphins.

NEW WORD	CLUE WORD OR IDEA	DEFINITION
jeopardize	people would lose their jobs	to expose to danger

Both Manuel and Kioko believe they are right about the dolphin issue. Which student do you think is right? Why?

As you read the viewpoints in this book, keep a list of the new words you come across. Write down what you think the words mean from their context. Check your definitions against those in a dictionary.

CHAPTER 1

PREFACE: Should Endangered Species Be Saved?

Which is more important: the survival of a hundred red squirrels or building a $200 million telescope that will enable scientists to learn valuable new information about outer space?

The citizens of Arizona are debating this very question. The answer is not easy to find. The state of Arizona wants to use Mount Graham as a site for a new telescope and space observatory. Promoters of the new development say it will be a great boon to human knowledge of life on earth, and will create jobs for many people. People who are concerned about the environment—how we treat the planet and the creatures that share it with us—do not agree, however. These people argue that the plan threatens to destroy the habitat or home of a certain species of red squirrel. Mount Graham is the only place in the world in which this particular kind of red squirrel lives. Environmentalists oppose the state's plan because they do not want to see the squirrels wiped out. They believe that even a species that seems unimportant may in the future prove to be the key to survival for other species—maybe even humans.

The Mount Graham controversy is part of a larger issue. The growing human population threatens the survival of more and more animal and plant species every year. Is this simply part of the natural development of life on earth, the law of evolution that says only the strongest survive? Or do humans have an attitude of carelessness toward other creatures? The viewpoints in this chapter debate whether humans should make an effort to save endangered species. As you read, you may see words you do not understand. Try to determine the meaning of these words by their use in context. Keep a list of words that are new to you.

VIEWPOINT 1 Endangered species should be saved

> **Editor's Note:** This viewpoint argues that humans should save endangered species. It states that people, plants, and animals all share a common life on earth. And, it argues, humans need nonhuman creatures to keep in touch with their own humanity and with nature. Without the influence of plants and animals, people would be less human.

Lost forever is the clue to the meaning of *irredeemable*. *Irredeemable* means *hopeless*. An *irredeemable* loss is one that cannot be recovered.

Many species of animals and plants today face extinction. They are in danger of disappearing altogether from the earth. Like the dinosaurs and the dodo bird, they will be lost forever.

People can and must prevent such an **irredeemable** loss. Why? Stephen Kellert, a forestry teacher at Yale University in Connecticut, gives this answer: "The health and well-being of wildlife and natural habitats are linked to human well-being and even survival." Many other environmentalists agree with Professor Kellert.

Protecting endangered species from extinction is good not just for those species, but for humans as well. Without the influence of wild creatures and wilderness areas, people would become more like their computers and less like human beings. They would soon long to hear a cricket sing or see geese flying overhead or smell the sweet dampness of a marsh in the morning.

The whole idea expressed by this paragraph hints at the meaning of *vitality*. *Energy* and *alive* especially point to its meaning. What does *vitality* mean?

Wildlife gives people a sense of beauty and gracefulness, of energy and **vitality**. People feel alive and relaxed around natural beauty. That is why they hang pictures of animals and plants in their homes and workplaces. That is also why they enjoy living in or visiting wilderness areas. Seeing, smelling, and touching the natural

© Rothco. Reprinted with permission.

environment reminds people that they are part of nature too, and so they feel more at home. When people feel at home, they enjoy life more.

Earth does not belong to humanity. It is home to nearly 30 million different kinds of creatures. People need to realize that they share the earth with all of these creatures. Humans, their pets and houseplants, the trees and flowers, even the housefly on the window and the mouse in the basement all breathe the same air.

People often hear about a building project that threatens the survival of some small, seemingly unimportant creature—a snail, for example. They may wonder why the loss of such a creature matters. Dartmouth College ecology professor Donella Meadows offers three reasons—all of which affect the good of humans.

One, the loss of a species may mean the loss of a valuable new product that would benefit humanity. Many newly discovered species of plants and animals have provided miracle medicines, foods, and other materials.

Two, every species is a link in a great chain of life. That chain needs every link in order to function. As an example, Meadows points out that if pesticides were to wipe out all the bees, humans would have to pollinate all of the fruit trees themselves. This task would be so expensive that it would **devastate** the agricultural economy. Much of the farming industry would be totally wiped out. And there would be no honey besides!

The third reason may be the most important. Each individual creature carries within its body's cells **irreplaceable** information about how to adapt to change and so how to survive. No other creature has that same information. When a whole species becomes extinct, the world loses a lot of "survival data." Such biological information could make the difference in the future survival of another species—including humans.

Many other reasons exist for saving endangered species of plants and animals. The few reasons discussed here show the importance for humans of saving endangered species.

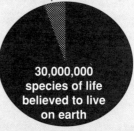

Preserving the Variety of Species

Only 1,700,000 species identified

30,000,000 species of life believed to live on earth

Humans cannot fairly judge which species can be allowed to become extinct. Too many species are unknown. Which ones will prove most useful?

Source: *Los Angeles Times*, May 13, 1990.

The farming industry would be totally wiped out is your clue here. It repeats the same idea as *would devastate the agricultural economy*. What, then, do you think *devastate* means?

You can determine the meaning of *irreplaceable* by breaking it down into its parts. The root word is *replace*. *Ir* means *not* and *able* means *able to be*. What does *replace* mean, then, with these syllables added? The ideas expressed in the paragraph can also help you determine the meaning of *irreplaceable*.

Why should endangered species be saved?

This viewpoint lists a number of reasons we should save endangered species. It says there are many other reasons for saving these species. What other reasons can you think of to save endangered species?

VIEWPOINT 2 Endangered species should not be saved

Editor's Note: This viewpoint argues that people should not try to save endangered species. Extinction is part of the natural course of development for life on earth. To prevent extinction is to stall this development and, in the long run, to harm humanity and other life, the author says.

The clue to the meaning of inexorable *is* no one escapes. Inexorable *means cannot be avoided.*

The capacity to think and the ability to think express like ideas. Like ideas can help you determine the meanings of unfamiliar words. What, then, does capacity *mean?*

It never stops is the clue to the meaning of perpetual. *The idea in the sentence containing* perpetual *is repeated in the following sentence. What does* perpetual *mean?*

The extinction of species of plants and animals is part of the natural course of life on our planet. Death is a necessary and **inexorable** part of life. No one escapes it. Just as individuals die to make room for new individuals, so species die out to make room for new species. This process is part of the natural development of life on earth.

This development, or evolution, has one basic law by which it works: Those species that can best adapt to their environment have the best chance of survival. As the environment changes, species either adapt to the new situation or die out.

The dinosaurs, for example, lived on earth for 160 million years. Then they died out. Some experts think they became extinct because they could not adapt to changing conditions in their environments. Other animals were able to adjust and survive.

The natural process of evolution made the human race possible. Less developed life-forms gradually gave way to more highly developed forms. Finally, one of those forms developed the **capacity** to think and to reason. That is, it became human. The ability to think greatly increased its chances for survival. Humans are here today because of that process of evolution that included the extinction of numerous species.

The evolutionary process is a **perpetual** one. It never stops and, in fact, is still going on today. When species become extinct, their extinction is simply a sign of continuing evolution.

Humans are changing the environment. Perhaps in a way they are speeding up evolution. Species that cannot adapt to the changes fast enough may die out. Others will flourish in the new environment.

Those who want to prevent the extinction of endangered species claim that humans are interfering with the course of nature. But these very people do not see that their own attempts to save endangered species may also interfere with the course of nature.

"FRANKLY, PROFESSOR, I THINK WE'RE THE ONLY ENDANGERED SPECIES AROUND HERE!"

© Rothco. Reprinted by permission.

Their **interventions**, however well intended, may in the long run work against them and their children.

Norman Levine, a former agriculture professor at the University of Illinois, believes that perhaps 95 percent of all species of plants and animals that have existed on earth are now extinct. He asks, would the earth or humankind be better off if these extinct species returned? No, he says. Would humans be able to peacefully coexist with great dinosaurs? Should scientists again introduce the smallpox virus into our world, after finally stamping it out in the late 1970s? No.

Humanity is struggling to survive in the world just as other species are. Humans at times need to change their environment in order to survive or flourish. This may endanger the existence of other species. Oftentimes that cannot be helped.

It may seem that humans have an unfair advantage in this struggle to survive. That advantage is their superior brains. But those brains were developed by the process of evolution to be just that: an advantage for survival. Those species that nature fits to survive will survive; those not fit will not survive. We must allow nature to take its course.

Look for clues to the meaning of *interventions* in the surrounding words and ideas. Whose *interventions* are being talked about here? What are these people doing? What does *interventions* mean?

Is extinction always a bad thing?

This viewpoint argues that death is a natural part of life. Do you agree with the author that people should not try to prevent the extinction of species of plants and animals? Why?

CRITICAL THINKING SKILL 1
Understanding Words in Context

The sentences below are adapted from the viewpoints in chapter 1. Try to determine the meaning of each highlighted word by considering its use in context. You will find four possible definitions of the highlighted word under each sentence. Choose the definition that is closest to your understanding of the word. Use a dictionary to see how well you understood the words in context.

1. The loss of a species may mean the loss of a valuable new product that would **benefit** humanity. Many newly discovered species of plants and animals have provided miracle medicines, foods, and other materials.

    a. destroy
    b. call
    c. attack
    d. help

2. If **pesticides** wipe out all the bees, humans will have to pollinate all the fruit trees themselves.

    a. crowds of people
    b. bad smells
    c. insect killers
    d. bright lights

3. This task would be so expensive that it would devastate the agricultural **economy**. Much of the farming industry would be totally wiped out.

    a. vegetable garden
    b. high technology
    c. business structure
    d. meeting room

4. Each creature carries information within its body's cells about how to adapt and survive. Such "survival data" may be lost forever if a species becomes extinct. This **biological** information may be another species's key to future survival.

    a. having to do with music
    b. having to do with food
    c. having to do with living things
    d. having to do with computers

5. Is this simply the natural course of development, the law of **evolution** that states that only the strongest will survive?

    a. survival
    b. circle
    c. war
    d. growth

CHAPTER 2

PREFACE: Do Humans Cause the Extinction of Other Species?

Earth's human population is rapidly increasing. At present, 5.5 billion people live on our planet. Zero Population Growth, a national organization that works to balance population and resources, estimates that world population may triple by the year 2050.

As the human population grows, it needs more food, shelter, water, and land. Humans move into wild areas that are home to many plants and animals. The human newcomers change the wild environment to make it support themselves. But in changing the environment, they also destroy the habitat and food supply of many nonhuman species residing there. These species' survival is then endangered.

Many people believe humanity is the main, if not the sole, cause of extinction for most animal and plant species. These people fear humans will wipe out so many species that human life itself may be endangered.

Others see humanity as the savior of endangered species. Humans have developed the means to preserve species and prevent their extinction. As human civilization expands and develops, people can either wipe out other species or preserve them. The viewpoints in this chapter debate which it should be.

As you read, try to determine the meaning of the highlighted words from their use in context.

VIEWPOINT 3 Humans cause the extinction of other species

Editor's Note: This viewpoint argues that humans are the biggest threat to the survival of other species. It lists four kinds of human behavior that have caused the extinction of other species throughout human history.

Enumerates sounds like the more familiar words *numeral* and *numerous*. Can you guess the meaning of *enumerates* from these clues and from its use in context?

What clues from the context provide the meaning of *decimated*? What does *decimated* mean?

"THERE WE WERE, ALMOST ALONE ON THE ICE, FACING TWENTY-FIVE POUNDS OF FEROCIOUS BABY SEAL, WITH ONLY OUR CLUBS FOR PROTECTION!"

© Renault/Rothco. Reprinted with permission.

The greatest threat to the survival of most animal and plant species is human beings. Paleontologists, scientists that study the past, have noted that everywhere humans have appeared many types of animals and plants have soon after disappeared.

Jared Diamond, a professor at the University of California at Los Angeles, **enumerates** four principal ways human beings cause the extinction of other species:

1) *Overhunting.* Humans overhunt when they kill species faster than the species can reproduce. Overhunting caused the extinction of many prehistoric mammals, including the woolly mammoth and the saber-toothed tiger. Today, hunting is still driving earth's largest animals toward extinction. Overhunting by humans has **decimated** the numbers of rhinoceroses, elephants, and whales. All of these are endangered species today.

2) *Introducing new species into a region where the species does not naturally live.* Bringing a new species into an area upsets the established balance of nature there. Almost without exception the new species destroys native species that have no natural defense against it or that cannot compete with it for food.

For example, in the fifteenth and sixteenth centuries, European explorers and traders brought goats with them to islands in the Pacific Ocean. Before that, these islands did not have goats. The goats had no natural enemies there and soon overran the islands. They ate up all the food and competed for territory once used by native plants and animals. Soon many native species were extinct.

3) *Destroying habitat.* Most species of animals and plants live in environments to which they are specially suited. They cannot live anywhere else. If their environment is changed, they will die out.

Rain forests are the most dramatic example of habitat destruction occurring today. Only 6 percent of the earth's surface is covered by rain forest. Yet more than half of all the world's species of plants and animals live in rain forests. Environmentalists predict that within sixty years most of earth's rain forests will have been destroyed by humans. Thousands, maybe millions, of species of birds, animals, and insects that live in the rain forest will be destroyed also.

4) *The domino effect.* A line of dominoes, no matter how long,

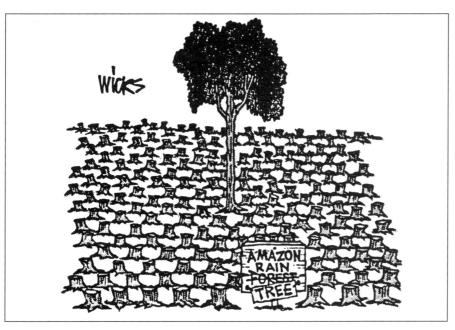

© Wicks/Rothco. Reprinted with permission.

can be toppled by simply knocking over the first one. This game is an **analogy** for what happens when even one species in a certain habitat is killed or driven off. Every creature depends on others for its existence and survival. Like a chain, if one link breaks, the whole chain no longer works.

For example, in Panama, the building of the Panama Canal killed off the native jaguar population. Then, the numbers of wild pigs and monkeys, its natural prey, increased. The monkeys ate so many bird eggs that some species of birds are endangered or extinct. The pigs tore up much of the plant growth that sheltered many smaller species of birds and animals. Without the jaguars, the delicate balance between species was upset. Other species were then threatened by extinction too.

At the current rate of extinction of species, half of all known species will be extinct by the year 2050. Unless something is done soon to **mitigate** this rush toward extinction, a great tragedy lies ahead for the earth. And humans will bear the blame.

An *analogy* compares something known to something unknown. If you know what a line of dominoes does, you can better understand the process that is set in motion by the extinction of a single species. The image of a chain used here is also an *analogy*. What does *analogy* mean?

From its use in context, what do you think *mitigate* means?

What is the biggest cause of extinction today?

This viewpoint lists four causes of species extinction for which humans are to blame. Which one do you think is the biggest or most common cause today? What might be done to reverse it?

VIEWPOINT 4 Humans prevent the extinction of other species

Editor's Note: This viewpoint argues that human beings offer the best hope of survival for endangered species. The viewpoint discusses some of the techniques that scientists use to help endangered species increase their numbers. Without this human help, many species would soon become extinct.

The explanation of the scientific process described in the text is the clue to the meaning of *fertilize*. What does *fertilize* mean?

Dozens of animal and plant species face extinction today. For most of these, the only chance to escape extinction comes from human beings. Human science, technology, and management skills are assuring a future for many species.

For example, scientists have developed a technique that enables a woman who is unable to become pregnant naturally to have a child. By this technique, just as in the natural course, a sperm cell from a male must penetrate an egg cell from a female. Only then can the egg cell develop into a child. This penetration normally takes place within the female's body. But scientists can now **fertilize** a woman's egg cell in a laboratory test tube. The fertilized egg, called an embryo, is then implanted in the woman's body to develop naturally. The embryo can even be placed in the body of a woman other than the woman from whom the egg was taken. This technique is called in-vitro ("in a glass") fertilization. This technique is also helping endangered animal species have babies and increase their numbers. Thus, these species can better avoid extinction.

Scientists have produced embryos by this technique for many endangered species. They keep the embryos frozen to use later if a

Species survival plans (SSPs) have saved many species from extinction. Zoos usually run these programs. This chart shows the gains a number of endangered species have made with the help of humans.

Species	Number in Wild Before SSPs	Numbers in Wild 1988 After SSPs
Kingfisher	26	46
Rail	10	95
Black-footed ferret	17	125
Arabian Oryx	0	323
Przewalski's horse	0	550
Golden lion tamarin	90	600
Pere David's deer	0	1,500
NeNe (Hawaiian goose)	30	2,050
European bison	0	2,500

species should become extinct in the wild.

Most of the time, though, zoos use these frozen embryos to produce animals for other zoos. For example, in 1983, the Cincinnati Zoo sent the frozen embryo of a rare and beautiful African antelope called a bongo to the Los Angeles Zoo. Zoo scientists in Los Angeles implanted the embryo in a female eland, another species of African antelope. The eland later gave birth to a baby bongo.

Many zoos have **compiled** a stock of frozen embryos from every animal in their collections. Scientists call these stocks "frozen zoos." Betty Dresser is the director of the Cincinnati Zoo's Center for Reproduction of Endangered Species. She has stated, "If someone had done this with dinosaurs, we could have dinosaurs today."

> To determine the meaning of *compiled* here, you have to understand the sense of the whole paragraph. What do you think *compiled* means?

Most zoos today also have enacted "species survival plans" to prevent extinction of endangered species. These plans include in-vitro fertilization, frozen embryos, and keeping track of species in the wild. Today, species survival plans are helping fifty-three endangered species survive.

One success story of human **intervention** to save an endangered species concerns the black-footed ferret. A ferret is a weasel-like animal that lives in the western United States. Or rather, it used to live there. No black-footed ferrets are known to be living in the wild anymore.

> *Intervention* was used in viewpoint 2. What does *intervention* mean?

In the 1980s, wildlife scientists realized that there was only one family of black-footed ferrets left in the wild. The scientists became very concerned when, one by one, the ferrets died without producing any **offspring**. Finally, there were just two left, a male and a female. The scientists decided to act. They managed to capture the ferrets and move them to a special laboratory where they mated and had babies. The scientists have spent years studying, protecting, and nurturing the ferrets in their laboratory. Now there are enough ferrets to allow some of them to return to the wild and multiply naturally.

> From its context, what do you think *offspring* means?

With the help of humans, endangered species may avoid being lost forever. The earth and all its inhabitants—especially human beings—will be the richer for it.

Are frozen zoos a good idea?

What do you think about the idea of "frozen zoos"? Would you like to be able to see a dinosaur in a zoo, even though they are now extinct in the wild? Do you think people five hundred years from now would be glad to be able to see an elephant, even if elephants only live in zoos by then?

CRITICAL THINKING SKILL 2
Using New Words in Sentences

enumerates
decimated
analogy
mitigate
fertilize
compiled
intervention
offspring

The words highlighted in viewpoints 3 and 4 are listed at the left. The short paragraphs below are each missing a word. Choose the word from the list that best completes each paragraph. You should be able to identify the appropriate word from its use in context.

1. Scientists have learned how to reproduce species in a laboratory. They can _____ an egg cell outside of the female's body and then put it back into her body to develop.

2. The image of a chain is an _____ for the relationship of all living things in the ecology. A chain is made up of individual links tied together. If even one link breaks, the chain breaks and does not work. So it is with the ecology.

3. Humans have taken over much of the habitat of wild animals. Many animals have nowhere safe to raise their young. Pollution caused by humans is another thing that makes survival difficult for the _____.

4. In viewpoint 3, Professor Jared Diamond _____ the ways that humans have driven other species to extinction. His list has four items.

5. Too many species are headed toward extinction. At the current rate of extinction, half of all known species will be extinct by the middle of the twenty-first century. Something must be done to _____ this torrent of doomed species.

6. Black-footed ferrets and other endangered species were almost extinct before humans stepped in to help. Many species have been saved because of human _____.

7. Illegal hunting of big game animals in Africa has _____ their numbers. Very few rhinoceroses, elephants, and bongos remain in the wild now.

8. Scientists all over the world are gathering data on various species of plants and animals. These scientists have _____ the largest amount of biological data in history. These data banks may save many endangered species from extinction.

CHAPTER 3

PREFACE: Does Hunting Threaten Species?

Seven percent of the American population—nearly 16 million people—hunt for sport. They kill about 200 million birds and animals annually. These figures from *U.S. News & World Report* magazine show two things: Hunting is a popular sport in America, and it has an obvious impact on wildlife.

Whether this impact is good or bad is a matter of debate. Hunters hold a positive view of their sport. For them, it is simply part of the natural order of life and death in nature. They claim that most of the prey they kill each year would have died from starvation or other causes anyway. At the same time, they argue, hunting provides enjoyment and appreciation of nature for millions of people.

Environmentalist and animal rights groups protest that hunting is unnecessary killing. It benefits no one, they say, neither the animal nor the hunter. Indeed, these groups argue, hunting harms the ecology. It upsets the natural balance among species in the wild. They blame pro-hunting game managers for this problem. The managers preserve game animals at the expense of their natural enemies and nongame animals, the groups say.

Is hunting a natural harvest of surplus animals? Or is it the slaughter of helpless, endangered wildlife that upsets the ecology? The viewpoints in this chapter argue both sides of the debate. Watch for the highlighted words and any other unfamiliar words in the text. Try to understand their meanings from the way they are used in context.

VIEWPOINT 5 Hunting threatens species

Editor's Note: This viewpoint argues that hunting endangers many wild species. The species endangered by hunters are not the ones hunted. They are the many other species whose habitats or prey are destroyed to make room for game animals. The author states that wildlife management people must be concerned for the ecology as a whole and not just for preserving game for hunters.

Hunting threatens many species of wildlife. This may seem obvious, since the goal of hunting is to kill the prey, or "game" animal. But hunting endangers nongame animals and plants as well as game animals.

Many hunters consider themselves conservationists concerned with maintaining wild animal populations and habitats. But they choose to conserve only certain species at the expense of others. This practice endangers the ecology.

Hunters and hunting organizations make sure that the animals they want to hunt are plentiful in the wild. They do this by influencing government policies on wildlife management. They pressure wildlife managers to create habitats and conditions favored by game animals.

For example, in order to keep an **abundant** deer population for hunters, wildlife managers create clearings in the forest. When trees and larger shrubs and plants are cleared away, new plants can grow up in their place. Deer feed on the shoots of new plants, so such clearings provide more food for deer than a mature forest does. Thus the deer population increases in forests with cleared areas. This is a

The clues to the meaning of *abundant* lie in the information given by the paragraph. If the deer population increases, what is an *abundant* population?

© Bender/Rothco. Reprinted with permission.

22 JUNIORS

boon for hunters, but it is bad news for many other species of plants and animals that live in the forest.

The government first established wildlife management programs because hunters and hunting organizations wanted such programs. Hunters wanted to make sure they would always have game animals to hunt.

Wildlife management did prevent many species of animals from being hunted to extinction. But today the wildlife management situation has changed. Scientists recognize that every wild creature is important to maintaining the balance of nature. Destroying some species or their habitats for the sake of increasing other species causes imbalance by lessening the **diversity** of species.

Hunters are not concerned with preserving a diversity of wildlife. They only care about preserving their intended prey. Wildlife management that caters to the needs and desires of hunters creates surpluses of deer, elk, and antelope. At the same time, it reduces the populations of nongame animals that compete for food and habitat with game animals. Hunters also view the game animals' natural enemies, like lions, wolves, and bears, as competition. So they destroy many of these animals in the process of trying to keep a surplus of game.

The state of Alaska, for example, once started a wolf **eradication** program to create a surplus of moose and caribou for hunters. State officials knew that wiping out the wolves would allow its natural prey to increase.

The Arizona Fish and Game Commission has used poison to kill animals that prey on pronghorn antelope. The antelope are also a favorite prey of hunters. The poison, however, also killed pets as well as animals that feed on the dead bodies of other animals. These **scavengers** are especially important for a balanced ecology. They help keep the environment clean.

Managing wildlife solely for hunting endangers many other species. The existence of every species must be protected if life on earth is to flourish. Many species are endangered today. Stopping hunting would be one of the easiest ways to save wildlife. Wildlife management practices must begin to focus on the whole environment, not on keeping hunters happy.

"Why don't they thin their own damn herd?"

What do you think *diversity* means, judging from the context in which it is used?

What clues tell you the meaning of *eradication*? What does *eradication* mean?

The text tells you what *scavengers* means. Can you find the definition? What does *scavengers* mean?

> Does hunting endanger some species?
>
> Hunters work to preserve and multiply game species and their habitats. What does the viewpoint say is wrong with this? How does such a practice endanger some species?

VIEWPOINT 6 Hunting saves species

Editor's Note: This viewpoint argues that hunters appreciate nature and wildlife. Hunters have made wildlife conservation possible with the license fees they pay. Without this income to fund management programs, the author says, many species would have become extinct by now.

What do you think contradiction *means, judging from its use in context?*

If it were not for hunters, more animal and plant species today would be endangered or even extinct. Because hunters kill animals, this sounds like a **contradiction**, but it is not. People who hunt for sport want to be able to continue hunting year after year. They must therefore ensure that the animals they stalk are plentiful. So, hunters have an interest in preventing these animals from becoming endangered species.

In fact, game animals—that is, animals that are popular as the prey of hunters—are among the most secure from extinction. Why? Because hunters, trappers, and anglers pay fees that governments use to maintain game populations and habitats.

In Africa, for example, hunting fees provide funds to preserve and manage big game animals and their natural environments. Most African nations are too poor to provide protection for their native animal populations. Poachers kill animals for a profit without paying any taxes or fees back to the government. Native farmers clear away forests and grasslands for agriculture without regard for wildlife. Without hunting fees, many endangered African species would soon become extinct.

What clues in the context point to the meaning of imperiled*? What does* imperiled *mean?*

The situation in America used to be like that in Africa. In the early years of the twentieth century, U.S. government officials despaired of the survival of many wild species. The numbers of many native animals were dwindling at an alarming rate. Hunting organizations urged the government to establish wildlife management programs to preserve these **imperiled** creatures.

Since Congress passed the Federal Aid in Fish and Wildlife Restoration Act in 1938, hunting and fishing fees have helped protect many species of wildlife. This money has built and managed wildlife refuges, constructed or restored more than three hundred lakes, and relocated endangered species to new homes. The graceful trumpeter swan, the colorful wood duck, and the shy wild turkey were nearly extinct at one time. Their numbers have all risen. These species have benefited from hunting and fishing fees.

What is the meaning of ban, *as used here?*

Hunters and hunting organizations lead the fight to organize and fund state conservation departments. To try to **ban** hunting, trapping, or fishing threatens a major resource for America's

Since 1938, state fish and wildlife agencies have used hunting and fishing license fees and special taxes under the Federal Aid in Fish and Wildlife Restoration Acts to:

Source: Archery World

- Acquire, develop, or manage 2,900 wildlife refuges and management areas totaling nearly 40 million acres.
- Construct or restore more than 300 lakes for fish and wildlife totaling 35,000 acres.
- Trap and move to new habitat more than 50,000 deer, 16,000 antelope, 2,000 elk, 1,000 wild sheep, 18,000 fur-bearing animals, 20,000 wild turkeys, 22,000 waterfowl, and 130,000 quail.
- Conduct studies on wildlife habitat needs, diseases, population growth or decline, etc.
- Help hundreds of thousands of landowners with wildlife habitat improvement projects.
- Educate the public about conservation through school programs, television programs, and magazine articles.
- Protect both game and nongame animals by arresting people who break conservation laws.

conservation efforts. According to *Archery World* magazine, for example, 61 percent of the 13.5 million acres of land preserved for migrating birds in this country is due to hunting programs. These birds need to rest on their long yearly flights in wild areas undisturbed by humans. Without preserved areas, many birds would not survive the journey.

Many people who do not hunt for sport have a mistaken idea of hunters. These people **envision** hunters as persons who do not respect nature or the life of wild creatures, but who just want to cruelly kill them. Nothing could be more untrue. Hunters appreciate nature. They enjoy the *total* experience afforded by hunting: being outdoors, sitting around a campfire, and stalking game through the woods and meadows. If sport hunters only wanted to kill something, hunting would not attract many followers. That is because relatively few hunters actually bag the animal they are hunting. But they still go out year after year for the excitement and adventure. Sport hunters enjoy and respect nature—often more than nonhunters do.

What other word do you see in *envision*? Is it a clue to the meaning of *envision*? What does *envision* mean?

Have hunters been misunderstood?

This viewpoint states that most wildlife management money comes from hunting and fishing fees. Hunters have also helped organize wildlife management programs. Do you think, then, that nonhunters have misunderstood hunters? Have nonhunters unfairly described hunters as cruel and thoughtless? Why or why not?

CRITICAL THINKING SKILL 3
Recognizing and Defining Cognates

A cognate is a word that comes from a similar word. The cognate is usually a different part of speech than the word it is taken from. For example, the word *conserve* is a verb, an action word. It means "to avoid wasteful use" of something. To conserve water, for example, is not to waste it but to use it only as needed. The noun form of *conserve* is the cognate *conservation*. It means the act or idea of conserving something. So, when we do not waste water, we are practicing water *conservation*.

Recognizing cognates enables a reader to determine the meanings of many unfamiliar words. For example, if a reader sees the word *conservation* for the first time, but knows the meaning of *conserve*, the reader can determine *conservation*'s meaning. Being able to recognize cognates increases a reader's skill in understanding new words.

The following sentences use cognates of some of the words that were highlighted in the viewpoints of this chapter. In the blank space before each sentence write the highlighted word from which the cognate is taken. A list of these words appears below. Then determine the meaning of the cognate from the sentence's context and write it down.

Some highlighted words from this chapter are:

a) abundant   d) envision
b) diversity   e) imperiled
c) scavengers   f) ban

___ 1. The life of a mountain lion or timber wolf is a *perilous* one. Hunters and ranchers often hunt these animals because they attack livestock and game animals.

   perilous means:

___ 2. If wildlife managers protect only a few different species, other species will die out. The lack of variety this would cause would threaten the future of life on earth. Earth needs a *diverse* number of species to survive.

   diverse means:

___ 3. In the 1800s, settlers *banished* the Indians from their lands to reservations. In the same way, humans are *banishing* the wolf and grizzly bear from their territory.

   banish means:

CHAPTER

PREFACE: How Can Elephants Be Saved from Extinction?

A bearskin rug, a leopard-skin coat, and a delicately carved ivory statue have two things in common. One, all three are beautiful treasures to possess. And two, all three can be obtained only by killing a beautiful animal.

Human beings have desired and possessed these things for centuries. But many people today are questioning whether it is right to kill animals just to provide luxury for humans. These people believe it is especially wrong to slaughter so many of these animals that they are threatened with extinction.

On the other side of this issue are people who believe that human beings are more important than plants and animals. These people do not believe killing animals for the sake of human needs is wrong.

The African elephant is just one species threatened by humanity's desire to own beautiful things—in this case, ivory. In October 1989, a group of nations concerned about endangered species met to discuss the threat of extinction for African elephants. They voted to make ivory trading illegal in their countries. They hoped this would save elephants. Some of the nations voted against this, however.

The following chapter debates the effectiveness of banning the ivory trade as a way of saving elephants from extinction. As you read, note the highlighted words and any other unfamiliar words you see. Try to determine the words' meanings from their use in context.

VIEWPOINT 7 Banning the sale of ivory protects elephants

Editor's Note: This viewpoint argues that making the sale of ivory illegal is the most effective way to save elephants from extinction. People will not buy much ivory if it is illegal, so poachers will be unable to sell much. Less poaching means more elephants, says the author.

Within elongated *is a word that provides a clue to its meaning. Its use in context also can tell you its meaning. What does* elongated *mean?*

Can you determine the meaning of ironic *from its use in context? What do you think* ironic *means? (Hint: it has nothing to do with iron.)*

The African elephant is an awesome creature. It is the largest land mammal in the world. A full-grown male can stand thirteen feet high and weigh seven tons. A peculiar feature of the elephant is its tusks. Tusks are **elongated** upper teeth that continue to grow throughout the elephant's lifetime. Because an elephant can live to be eighty years old, tusks can grow very large. They can reach ten feet in length and weigh two hundred pounds. The elephant uses its tusks to dig for food, to move tree trunks and other heavy objects, and to defend itself. Its tusks help the elephant survive in the wild.

It is **ironic**, therefore, that its tusks have made the elephant an endangered species. Humans kill elephants by the thousands for their tusks, known as ivory.

Ivory is hard and shiny and a rich cream color. Many people highly value its beauty. They like to own carvings made from it. In the past, wealthy people liked to have the keys on their pianos and the balls for their billiard tables made from ivory. Objects made of ivory were a mark of wealth and distinction.

© Whittemore/Rothco. Reprinted with permission.

But the fact that elephants must die so that humans can display a few expensive and pretty **trinkets** means a dim future for the elephant. In fact, at the rate elephants are being slaughtered for ivory, their future is worse than dim. It does not exist. According to Susan S. Lieberman, the associate director of wildlife and environment for the Humane Society of the United States, one thousand elephants are killed every week—almost all of them illegally.

As recently as 1980, 1.3 million elephants roamed the plains and forests of Africa. Today, about one-third of that number remains.

The destruction of the elephant's habitat has partly caused this decline in the number of elephants. But, by far, the primary cause is the illegal killing of elephants. Elephant **poaching** even occurs within game preserves where the huge beasts are protected.

As further evidence of the danger of extinction for elephants, wildlife management officials point to the smaller size of the tusks that are now being sold by poachers. The size of the tusk indicates the age of the elephant. The older the elephant, the bigger the tusk. The average tusk sold for ivory in 1972 weighed about thirty pounds. By 1989, the average weight dropped to little more than six pounds.

A six-pound tusk comes from a young elephant, one too young to have yet **reproduced**. Poaching is driving the elephant toward extinction because it is killing off the younger elephants before they can have offspring and increase the herd.

In 1988, the Convention on International Trade in Endangered Species (or CITES, for short), an organization of environmentally concerned nations, met in Switzerland to discuss a ban on selling ivory. The nations voted to ban international ivory trading. Since then, the price of ivory has dropped by 50 percent and fewer elephants have died at the hands of poachers, says Diana E. McMeekin, a reporter for *Wildlife News*. This turn of events is the best proof that elephants can be saved from extinction simply by banning the sale of ivory.

> From its use in context, what does *trinkets* mean?
>
> What clues tell you the meaning of *poaching*? What does *poaching* mean?
>
> From its context, what do you think *reproduced* means?

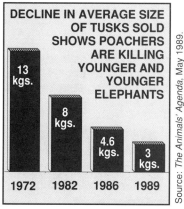

DECLINE IN AVERAGE SIZE OF TUSKS SOLD SHOWS POACHERS ARE KILLING YOUNGER AND YOUNGER ELEPHANTS

13 kgs. (1972) | 8 kgs. (1982) | 4.6 kgs. (1986) | 3 kgs. (1989)

Source: *The Animals' Agenda*, May 1989.

> Why does poaching threaten extinction for elephants?
>
> Why does killing very young elephants threaten the whole species with extinction? How does banning the sale of ivory help save elephants from extinction?

ENDANGERED SPECIES

VIEWPOINT 8   Banning the sale of ivory endangers elephants

> **Editor's Note:** The following viewpoint argues that making ivory illegal to buy and sell does not protect elephants but endangers them. It raises the price of ivory, says the author, which results in even more elephants killed for their tusks.

You can determine the meaning of confiscated *from its use in context. What does* confiscated *mean?*

Can you figure out the meaning of incentive *from its context? What do you think* incentive *means?*

POACHING IS OUT OF CONTROL IN EAST AND CENTRAL AFRICAN NATIONS

Southern African nations spend almost ten times more to prevent poaching than East and Central African nations.

Kenya 91%
Uganda 90%
Zaire 60%
Cen. Afr. Rep. 80%

Percentage of elephant population killed by poachers from 1980 to 1989.

Source: *The Animals' Agenda,* May 1989.

On July 18, 1989, the president of the East African nation of Kenya publicly burned twelve tons of elephant tusks. As ivory, the tusks would have sold for three million dollars. The ivory was **confiscated** from poachers. It was destroyed to demonstrate Kenya's commitment to saving elephants by banning the international ivory trade.

Banning the ivory trade, however, is the wrong approach to take to save elephants from extinction in Africa. In fact, such a ban will only drive them closer to the brink of extinction. Why?

Making ivory illegal to buy and sell makes it hard to get. That makes its price skyrocket. The **incentive** for poachers to sell illegal ivory is very high then, because they can make much more money. They poach even more elephants than when ivory trading is legal.

The nations of East and Central Africa have tried to save the elephants by placing them in wild animal preserves and national parks to protect them from poachers. But according to researcher Urs Kreuter, who works in the southern African nation of Zimbabwe, in the 1980s, East African poachers killed 56 percent of the elephants in national parks. (The poachers killed 78 percent of the elephant population outside of parks.) Living in a "protected" park evidently does not guarantee the animals safety from poachers.

Kreuter and others believe elephants are best protected by management programs like those operated by the southern African nations of Zimbabwe, Botswana, Namibia, and South Africa. In these nations, the elephant populations are not endangered. In fact, they are increasing. Why the difference between these nations and the nations of East and Central Africa?

Randy Simmons of the Competitive Institute in Washington, D.C., offers an answer. (The Competitive Institute studies economic and business issues.) Simmons says the southern African nations succeed because they handle the elephant problem like a business. They control the growth of elephant populations. They allow a certain number of elephants to be hunted by big-

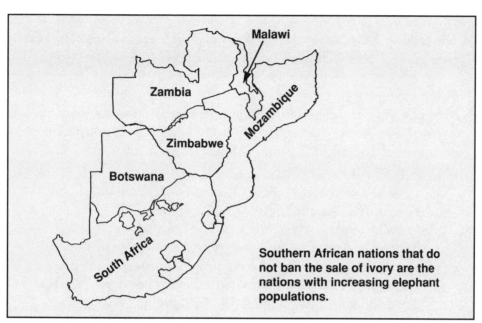

Southern African nations that do not ban the sale of ivory are the nations with increasing elephant populations.

Source: *Policy Review*, Fall 1989.

game hunters. They also allow the sale of ivory from the legally killed elephants. The profits from the sale of ivory and hunting permits return to the local people. These profits amount to many thousands of dollars for the people, so they are eager to care for the elephant herds.

Another major factor in the success of these management programs is a strict control of poaching. These countries spend as much as ten times more money to prevent poaching than East African nations do. The **penalty** for poaching in these countries is severe. In Zimbabwe, for example, poachers are shot on sight.

The strict laws against poaching and the people's interest in the elephant herds work together to protect elephants in these countries. This **cooperation** between government and people has proved to be a more effective way to save elephants from extinction than banning the ivory trade.

What does *penalty* mean? What clues helped you decide on its meaning?

The definition of *cooperation* is in the text. Can you identify it? What does *cooperation* mean?

Is the ivory trade the elephant's real enemy?

If ivory can be made available without threatening elephants with extinction, would ivory trading be all right? If so, what needs to be done to protect elephants from extinction? Do you think Zimbabwe's method of dealing with this problem is an acceptable one? Why or why not?

CRITICAL THINKING SKILL 4
Reading Comprehension

The following text is from a book on endangered species. The text uses some words that may be unfamiliar to you. Read the text and make a list of any words that are unfamiliar. Then try to determine the meanings of the unfamiliar words from their use in context. Next to each unfamiliar word on your list write down its meaning as you have determined it. Finally, write a short essay explaining your understanding of the text. In your essay, use your definitions of the unfamiliar words.

There are many practical reasons to preserve wildlife. We depend on trees and vegetation, most of them in forests, to recycle carbon dioxide into the oxygen we breathe. Wild plants and animals provide us with food, medicines, and commercial products from shampoo to floor wax. Hunting, trapping, and fishing are important wildlife industries that provide a livelihood for many people. New crops and livestock can be developed by crossing domestic strains with hardier wild varieties. And tourism to wilderness areas gives developing nations desperately needed income.

Yet there are other valid reasons for saving the plants, animals, birds, and fish of the world. Some people find the sheer beauty of nature a good reason. Others feel that killing living things for no reason is wasteful. Still others think it is arrogant for us to claim the whole world to do with as we please. Perhaps all of these reasons motivate the thousands of people who are working to save endangered species.

From *Endangered Species*
by Sunni Bloyd